Building an

UNCONVENTIONAL
BIOTECH

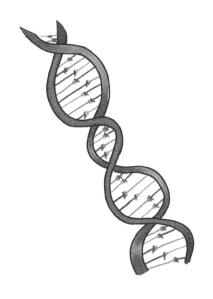

BLAISE LIPPA, PhD

Introduction

On February 19, 2009, the European Medicines Agency recommended that sales of the psoriasis drug Raptiva be suspended in Europe. Two days later, Canada Health advised that all marketing for the drug be ceased. In April, Genentech—the makers of Raptiva—announced a "phased voluntary withdrawal" of the drug. By June, it could no longer be obtained in the U.S.

Raptiva had been approved by the FDA in 2003 and spent six years on the market before its sudden retraction. The reason behind its removal was it caused a rare and deadly central nervous system disease known as progressive multifocal leukoencephalopathy (PML). The disease occurs when a common virus, the John Cunningham (JC) virus, is activated in the human body. At least 50% of the population lives with the JC virus inactive in their body, but only a select few have to worry about it becoming active. However, when the autoimmune system is weakened by disease or medication, the virus can enter the central nervous system and develop into PML.

At first, the connection between the drug and the deadly virus was unclear. Raptiva is what's known as an integrin-inhibitor drug. It works by inhibiting a specific receptor on white blood cells, keeping them from mistakenly overreacting and attacking the body. Because psoriasis is caused by an overactive immune system, Raptiva was able to effectively treat the condition by inhibiting that integrin on white blood cells. Yet its effectiveness at keeping the white blood cells from doing their job was the very thing that allowed the JC virus to gain a foothold in the body and unleash PML, usually resulting in death or severe disability.

The discontinuation of Raptiva was unpleasantly reminiscent of the failures of other integrin-based drugs—a promising solution with unexpected side effects.

Although integrins went undiscovered until the 1980s, they facilitate some of the most important processes within our bodies. An integrin is simply a receptor protein on a cell. Its role is to act as a connector, adhering the cell to the extracellular matrix in the body's tissue after it receives the correct signal. That adhesion is how white blood cells, which can travel through the body at speeds of 100 µm/s, are able to stop and make their way into damaged tissue at the correct point within the body. However, integrins aren't limited to white blood cells. There is a profusion of unique integrins in our various tissues, each with a specific job related to cell adhesion.

One of the most prominent researchers working during the initial period of integrin discovery was Dr. Timothy Springer. After receiving a PhD in Biology from Harvard, Springer completed a fellowship in Cambridge, England, then returned to Harvard Medical School as an assistant professor in 1977. Springer spent much of his early career investigating the biology of integrins, attempting to figure out how they work on a molecular level. His research was critical to our understanding of these proteins and how they control and regulate responses in the body.

As the research rolled out, the medical and pharmaceutical community began to realize the potential of integrins for the treatment of many diseases, including thrombosis/blood thinning and autoimmune diseases. In short order, the field saw an influx of money and integrin-inhibiting drugs began to work their way toward FDA approval. But soon after clinical trials began, complications arose.

As pharma companies labored their way through the arduous process of drug discovery and development, one thing became clear: the intravenous drugs were working; the oral drugs were not.

This unpleasant complication was hammered home by the failure of pharma companies to develop a safe oral drug for $\alpha IIb\beta 3$, an integrin involved in binding platelets together. Researchers began working with $\alpha IIb\beta 3$ for a very simple reason—when too many platelets join together, they can cause a clot. So it seemed logical that inhibiting the $\alpha IIb\beta 3$ integrin could be a straightforward solution for preventing blood clots.

Initially, the drugs developed to prevent clotting were successful.

In fact, there were several acute, intravenous drugs approved during that initial foray into the science of integrins. These drugs only stay in the body for a few hours, and they're still used today in thrombosis patients who've had a clotting episode.

But researchers and pharma executives were hungry to take the next step toward making integrin inhibitors a household name: the creation of an oral drug for chronic conditions. So scientists aimed to create a daily oral pill for people who were at risk for cardiac events or strokes. It would work in roughly the same manner as the intravenous drug, thinning the blood by inhibiting the αIIbβ3 integrin and preventing clotting.

The concept was tantalizing, and six different companies took these oral integrin drugs all the way to phase III clinical trials—where thousands of volunteers were given the drugs to test their efficacy. Yet, as is often the case in the pharmaceutical realm, the biology was more complicated than anticipated. Instead of reducing clotting, the drugs shockingly had the opposite effect—actually increasing mortality rates in patients.

In 2002, Eric Topol, the Chairman of Cardiovascular Medicine at the Cleveland Clinic, advised companies to stop testing the drugs. The risk for patients was simply too high for the potential reward. And since the Cleveland Clinic is one of the most well-known and respected hospitals for cardiovascular disease, Topol's warning carried serious weight.

After numerous defeats and warnings, it seemed useless for pharmaceutical companies to continue pursuing oral integrin drugs and sinking more resources into discovery and development. Like many technological advancements, the attempt to develop these drugs aligned almost perfectly with the Gartner Hype Cycle—early, overly-inflated expectations followed by deep disillusionment, which then eventually transitions into a slower, steady pace of innovation and discovery.

As it became clear that the oral drugs weren't working, and actually having the opposite intended effect, researchers began to abandon the field. Money dried up.

But not everyone had given up on integrins. Tim Springer wanted to learn more about what had gone wrong with the oral integrin drugs. As one of the researchers behind the initial discovery of integrins, he had a deeply rooted interest in figuring out why the drugs were doing the opposite of what logical reasoning predicted they should.

With the help of Albert Lin, a postdoc at Harvard who spearheaded the effort, Springer and his team were able to learn more about integrin conformations—a critical component of the efficacy of integrin-inhibitor drugs. Their key insight was that the oral drugs designed for inhibiting the integrins were mimicking their natural binding partner, paradoxically turning them 'on' and putting the integrin in an "activated" conformation. Instead of helping to thin blood, the activated integrins were priming patients for future clotting. The drugs meant to help the patients were actually ticking time bombs priming them for disaster once the drug concentrations dropped, usually in the middle of the night.

Once they realized where the drugs had gone wrong, Springer and his team focused on figuring out how to lock the integrin into a closed inactivated conformation to deactivate its tendency to adhere and cause clots. Lin was able to design a system to crystalize the integrin and solve X-ray structures of multiple bound drugs stuck to the integrin. Essentially, this is a 3D blueprint of the integrin showing how it works. This was the eureka moment—they found the answer for how to truly inhibit not just one integrin, but the whole family, and they were the only ones who knew!

Springer and Lin's ability to solve these crystal structures was the breakthrough needed to give oral integrin drugs a second chance. Except this time, it wouldn't be major pharmaceutical companies like Pfizer or Genentech handling the research and discovery. It would be a small, startup biotech company, Morphic Therapeutic.

PART I

Founding Story

CHAPTER 1

To Discover New Drugs, Biotech Is Best

When most people think of the pharma industry, they picture "big pharma"—massive corporations producing an endless array of drugs for everything from cancer to Crohn's Disease. What they don't see are the thousands of smaller biotech companies, funded by venture capitalists, that actually develop the majority of the molecules and drugs that end up treating patients.

While the pharmaceutical industry goes back a century, Genentech, the first publicly owned biotech company, was founded in 1976. Now a large player in the pharma industry, the company was born from the humblest of beginnings. Its founders, Robert Swanson and Herbert Boyer, each invested $500 to start the company. But by 1979, Genentech had accomplished an astounding feat—producing the human version of insulin, which in the past had to be obtained from pig and cow pancreases. Due to their groundbreaking work, the company went public in 1980 and raised $35 million. After a string of successes over the next few decades, Genentech was acquired by Roche in 2009 for $46.8 billion.

Genentech's rise was revolutionary at the time, but its origin story is much more commonplace today. Many small biotechs aim to get enough capital funding to establish proof of concept and then partner with (or be acquired by) a larger company to push their drugs across the finish line and into production. Though of course, few reach the heights of success accomplished by Genentech.

In addition to Genentech's success, two other events in the 1980s made the rise of the modern biotech industry possible. First, in the case of Diamond v. Chakrabarty, the Supreme Court ruled that genetically modified organisms could, in fact, be patented. In addition, Congress passed the Bayh-Dole Act, which allowed the recipients of federal research funds to obtain patents. Together, they helped open the gates to the current biotech cycle: research in academia, discovery in smaller biotechs, and development and commercialization with the help of large pharma companies.

Those laws are why much of the high-risk, basic research that happens in the pharmaceutical world today starts in academia. Companies and their investors simply can't spend 10+ years researching a target and using up millions of funding dollars without any end in sight, in part because global patent laws only give companies 20 years of protection from discovery. During this patent time period, they are permitted to have a monopoly on a new drug, which can be quite lucrative. Once the patent expires though, the discovery is available to all—and prices plummet. Due to the very high cost and risk of discovering a new drug, this profit incentive is needed to spur investment. Yet, companies can never rest after a success since the clock to patent expiration is always ticking.

But the timelines for university research labs are much more forgiving. They have access to government grants and students, and their goal is to learn more about the basic science while training scientists, instead of trying to make money.

In essence, academic labs can take risks that businesses could never justify, and in doing so, help seed the vibrant drug discovery ecosystem we have today.

Once a significant development has been discovered in academia, a smaller biotech company is often created in order to take the concept and run with it. At this stage, the concept is still perceived as too risky for a major pharma company to engage with. Yet a small startup can focus all its energy and resources on the project by continuing the research and taking risks in the drug discovery process until it looks like there's real potential to develop a drug. Once this happens, a larger pharma company will often swoop in and acquire the startup, or at least license the drug candidate.

This system works fairly well because established pharma companies are much better at performing large clinical trials, and subsequently commercializing the drug. They aren't as successful at early-stage discovery research. Which is strange, when you first think about it. These big pharmaceutical companies have the resources and capabilities to handle a large number of failures. They could constantly be financing drug discovery. Yet to this day, for a variety of reasons, their research groups have relatively low productivity.

All of this isn't to say that large drug companies can't do anything right. They're extremely proficient at certain portions of bringing a drug to market. However, there's a reason why someone like Tim Springer wanted to found a small startup and expand on his lab's integrin research—big companies are often too distracted and risk-averse to be trusted to see a project all the way through.

Anyone who works in drug discovery knows the job entails handling a series of highs and lows. A drug that seems to be on the road to success will end up hitting some unforeseen roadblock, like toxicity, for a reason no one could have predicted. That's par for the course. There's always an element of surprise when a drug doesn't work out, but it isn't necessarily unexpected. In fact, there's roughly a 50/50 chance that a scientist will work their entire career and never even touch something that becomes a drug. Even if the drug does get into human clinical trials, the failure rate at that point is still about 90%.

People in the biotech industry have to be at peace with failure since they're going to dedicate 30+ years of their lives to an industry where failure is essentially the norm. That doesn't mean losing all hope or giving into fatalism, however.

Tim and his team didn't want to see oral integrin drugs shelved at the first sign of problems. Yes, there had been failures in the past when it came to producing oral integrin drugs. But with the new research from his labs, Springer felt the time was right to try again.

The odds of success had to be top-of-mind during the summer of 2015 as Springer's company, Morphic Therapeutic, began coming together. Everyone on the initial team had plenty of experience in the field and knew that taking a chance on integrin drugs would be difficult. If it were easy, it would have worked the first time around.

CHAPTER 2

Reckoning With Biotech Roles

On March 5, 2015, I was laid off for the first time in my life. I had a feeling it was coming when the CEO of my company at the time, Cubist, announced that we were being acquired by Merck—but that knowledge wasn't much help when the shoe finally dropped.

Merck bought Cubist on Monday, December 8, 2014. The Friday before, all was quiet and normal at work. Yet that weekend, I decided to check Cubist's stock price while I was sitting in the stands at my son's hockey game. It had spiked to almost $100 per share—clearly, something was about to happen! The acquisition was announced as everyone rolled into work on Monday.

Initially, my team was hopeful that we wouldn't be laid off. The messages we heard from Merck's leadership were positive. "You all have so much knowledge in this area. It would be silly for us to fire everyone and just take the products." They even gave us Rubik's Cubes with 'Merck' printed on one side and 'Cubist' on the other. I still have mine in my office at Morphic. Though this all sounded positive, there were never any guarantees and they were careful not to lie.

Of course, three short months later, they called our discovery team into the conference room to tell us we were all being laid off. The experience was slightly surreal, and most of us were sad. Cubist was a special company that always treated us well and felt like a family. Now it was all over. They handed everyone a blue packet with a jumble of legal documents about severance and an agreement not to sue. Still, I was fortunate. Merck paid a hefty sum of $9.5 billion for Cubist, gave us a generous severance, and it was considered a very productive exit in the venture world.

The terms of the buyout also left me some breathing room. Cubist had laid off my entire department, triggering the WARN Act—a law that stipulates an extra two months of pay for employees who are part of mass layoffs. I officially had 60 days to figure out my next step.

When looking for a job, the biggest fear is not finding one. The second biggest fear is picking the wrong job. Coming from Cubist, I had a positive story to tell prospective employers. But I knew if I made the wrong move, my next story wouldn't be as rosy. I found myself at a crossroads and was very cognizant of the importance of taking the right path forward.

Luckily, I had several potential options on the table, including a fledgling biotech startup named after a rock.

My journey to Morphic began in the zig-zag way so many opportunities do. Several years earlier, I met a computational scientist named Mark Murcko at a scientific conference who happened to be involved with a number of smaller biotech companies. Once I saw the writing on the wall at Cubist, I began looking extensively through my LinkedIn contacts. I came across Mark's name and decided to reach out. I didn't want to come right out and ask for a job, of course. I just wanted to talk and get some advice.

Mark asked me to call him during his morning commute (at 6 a.m!), so while he drove, we talked about my experience and interests. After the call, he connected me with Kevin Bitterman—a venture partner at Polaris. Kevin looks extremely youthful, but he has phenomenal people skills and a PhD in biology from Harvard. He's one of those people who naturally excel at pulling others in and making connections, and his scientific background allows him to make informed judgment calls about the right science and the right person to fill a role.

I got on the phone with Kevin, and he described Morphic in fairly general terms. But from the moment I started talking to Kevin about the Head of Chemistry role, it struck me as being an exciting opportunity. I liked the target perspective. I could tell they had something unique with the crystal structures and a partnership that allowed them to use state of the art 3D computer modeling.

The only thing that initially gave me pause was the size of the company. At the time of my interviews, the company consisted of a handful of people working to start something from scratch. The road ahead would be long, with no guarantees. In biotech, it's risky to step into a role with so many uncertainties and such a long timeline to "success." For Morphic, that would mean bringing a drug to market or being bought by another company.

If I had been coming straight from my first job at Pfizer, I would have turned it down. But my experience at Cubist had shown me the long-term potential of a startup taking greater risks. I knew there was an opportunity waiting at Morphic that I might not find again.

Kevin and I followed up in person to continue our conversation. It was all very informal. Little did I know, this was the first of many meetings that would take the place of a traditional interview. Since then, I've come to find that formalities are uncommon in small companies. Those first few hires are extremely important, and if interviewers are doing their job, they're going to try to really get to know you over time.

Eventually, Kevin connected me to Tim—Morphic's founder. The day I met him was right after a farewell Cinco de Mayo pool party with Cubist colleagues, and I had two other interviews lined up in a row. I had no idea if any of them were going to work out, but I spent the day before doing as much research on integrins and reading as many of Tim's papers as I could.

I wasn't entirely sure what to expect from him.

Here's the thing about famous professors: they're always incredibly intelligent, but they often have monster egos to go along with their brainpower. In the past, I've met academics who were certain they knew everything about drug discovery, yet they lacked experience working for a company. That leaves a knowledge gap many aren't ready to admit to.

Tim is just the opposite. You can tell he's confident in himself— as he should be—but he's flexible and interested in exploring other people's viewpoints. More importantly, he has a wealth of experience working both in academia and in the business world as an accomplished investor and entrepreneur.

Tim founded his first company, LeukoSite, back in 1993. By 1998, the company had gone public, merging with Millennium Pharmaceuticals a year later. All of Millennium's drugs wound up coming from LeukoSite, and the company was valued at $3 billion by 2001. Eventually, it was bought by Takeda—the largest pharmaceutical company in Asia.

Subsequently, Tim went on to be one of the most successful Angel investors of all time. As an entrepreneurial professor, he recognized exciting research from colleagues at a very early stage. He then invested his own money to 'seed' this research, leading to companies such as Moderna and Editas. As the companies grew, so did Tim's stake, to the tune of hundreds of millions of dollars.

Despite this wealth, Tim often wears blue jeans and a t-shirt, bikes to work (when he isn't driving an old minivan) and is known to shop at Costco for a particular type of oranges. He is a genuinely nice guy who is approachable and dreams of establishing a perpetual scientific institute, rather than flying private jets.

In 2014, Tim himself founded his second company, Scholar Rock, which focuses on the discovery and development of niche activators that target disease-carrying proteins. The company's name reflects another of Tim's interests—geology. Actual scholar rocks form through erosion and wear, creating unique patterns and holes that often resemble proteins. During this second foray into entrepreneurship, Tim worked with the investment firm Polaris. Every so often, the company would hold founder retreats to bring entrepreneurs together to explore new ideas for companies.

It was during one of those retreats back in 2014 that Tim decided there was finally an opportunity for a company to develop oral integrin drugs. Up to that point, the field hadn't been promising due to the failures discussed, but his lab's crystal structures and insights had provided the spark to form a nascent company.

With the idea in place, Tim, Kevin, and Albert began building a team. The new company, Morphic Rock Therapeutic, would also have geologic connotations. Metamorphic rocks are formed when an existing rock is subjected to heat and pressure. A chemical change occurs within the structure, creating a new type of rock. While few adults know what metamorphic rock is, it's standard curriculum for every 3rd grader in the US.

The formation of metamorphic rock reminded Tim of the changing conformations of integrins. At Morphic, the hope was that years of research in the Springer Lab would create something new—and just as incredible.

CHAPTER 3

Designing A Biotech,
From People To Pipettes

Morphic Therapeutic didn't officially form until June 1, 2015, and it began without a lab or an office. The seed funding consisted of a combined $6.65 million directly from Tim and Polaris Partners. The company also had two partnerships: one with ChemPartner, a research organization located in Shanghai, and the other with Schrödinger—a chemical software company in New York City. Aside from the money and the partnerships, the company was a blank slate.

Our initial team was a small group that summer and was still missing many of the people who would become key members. But the original six were Albert Lin (employee number one), Praveen Tipirneni (our CEO—more on him later), Kris Hahn and Eric Humphries (two chemists I'd worked with at Cubist), Terry Moy (our first biologist, also extracted from Cubist), and myself. Bob Farrell, our head of operations, began as a consultant but would sign on full time a couple months later. Together, we began working to build the foundation of what we hoped would become a successful biotech startup.

You'd be forgiven for reading the word "startup" and imagining the six of us flying into San Francisco International to set up shop in Silicon Valley. But while the right zip code in California is a necessity for many tech companies, the beating heart of the life sciences startup scene is actually Boston, Massachusetts. Or, to be more specific, Cambridge—the city within the Boston metropolitan area that's home to Harvard and MIT.

Cambridge is the preeminent hub for biotech activity, but it also happens to be a pretty desirable place to live and work even for those without a PhD. And as we came to find out, calling the market for lab space in Cambridge "hot" would be a bit of an understatement. We were all working from home at the time, trying to find an office building for a reasonable price. It didn't take us long before we realized we'd have to get creative with our search.

Eventually, we found an opening in AstraZeneca's Biohub campus in Waltham, MA, which is about 20 minutes away from the heart of Boston. The location may not have been quite as desirable, but the office is situated in a biopharmaceutical campus and came at less than half the cost of what we'd found in Cambridge. And much less than half the driving commute for many of us!

In the early days, expenses were very much on everyone's mind. Keeping us in check was Bob Farrell, our head of operations. He basically had responsibility for everything other than science - Finance, HR, IT, building permits, you name it. Bob is a native Bostonian with a quick wit and a skill for hard negotiating the best prices. At the same time, you can count on him to be completely honest. He had worked with Kevin previously, and when he became free, Kevin grabbed him for our team. Before moving in to our office space, Bob organized trips to pick out second-hand lab equipment and negotiate licenses for software. He's the reason we still never fly business class, even on 15-hour flights to China. (Thanks Bob!)

But in those early days, a money-saving mindset was necessary. We quickly realized that the funding wouldn't last as long as we'd imagined—especially when we were trying to set up a complex and expensive biotech lab. There was one point later on when Bob had to remind us we only had about a month's worth of money left.

Despite the stress and pressure to get Morphic up and running as quickly as possible, the first few months were some of the most exciting times of my career. Starting a company from scratch was a new challenge since all of us came from larger companies, and the feeling of being involved in every decision was exhilarating. We all pitched in and helped where we could. I vividly remember visiting warehouse after warehouse, picking through equipment to find what we needed to construct a working lab.

The mood when we finally moved into the AstraZeneca building in August 2015 could be best described as pure excitement. We finally had space to meet and start working on the science. This was essential because even with all the tasks we had been doing, it was difficult to build camaraderie while working from home. Finally seeing people every day and getting to know them was a huge step in solidifying the team and building our own culture.

Still, it took some time for everyone to adjust to our new working arrangements and space. The AstraZeneca campus is not a typical first location for a startup. It isn't a 20x20 office at a coworking space. The building is gorgeous, with huge windows, bright open labs, and offices full of modern furniture. Interestingly, being located on the campus of an international pharmaceutical company only further drove home the feeling that our company culture would be much different than other biotech startups. We had a lot to live up to.

A hint of this dichotomy between startup and big pharma happened on our first day. We immediately had a 4-hour long orientation to learn about the building, the security, and the operations. Each office received a key, and at the end of our orientation, we realized one of the keys was missing. So, I went down to the security desk and asked them for the missing key. What followed was a 20-minute conversation among the security officers about the various issues that might arise from giving me the key. Should they have me fill out a form? Would that be enough? What was the protocol for missing keys? In the end, they just couldn't do it. We had to wait another day to get the key.

I found it symptomatic of the differences between a smaller biotech company like our own and a massive corporation with a large internal bureaucracy. It's impossible to check an ID and hand over a key—not when there are forms to fill out. While I laughed at the time, I soon realized this was the first of many experiences where the Morphic mindset would begin bucking the traditional biotech system.

Once all our keys were in hand and our lab was set up, we started honing in on hiring. We knew that our upcoming Series A funding round would have to be a success, so we began pushing harder to bring in people who would help build upon the newly solidified foundation of the company.

PART II

Building A Team

CHAPTER 4

Our Chief Concern: Finding A CSO

Lab space wasn't the only tough thing to find in Cambridge. Getting the necessary scientists to *fill* that lab space was also a challenge. The thriving biotech scene in Boston does have some similarities to its Silicon Valley counterpart, one of them being the wealth of opportunities available to talented individuals. This demand for highly-skilled employees has led to something of an arms race in benefits. Some startups have begun offering student loan repayment, unlimited vacation plans, sabbaticals, fitness reimbursements, and fully-stocked fridges. As you can imagine, talented candidates get snapped up quickly.

But we soon discovered that with an undeterred persistence and extensive networking, it was possible to find the people we needed.

While we didn't offer quite as many perks as more established companies in the area, we did have snack delivery on Mondays, bagels on Wednesdays, happy hour at the end of the week, and a positive, tight-knit company culture. Of course, there was also the tantalizing offer of stock in a platform company with a historically successful founder and multiple shots at bringing a drug to market. That type of opportunity tends to outweigh a gym membership in the long run.

Some of our new hires wanted the opportunity to prove themselves outside of their old positions. Others were casualties of mergers or company closures. Everyone had their own story of how they came to Morphic, but together, they formed the nucleus of the company we hoped to build.

In the early stages, each and every hire we made was critical. One person could have significantly changed our outcome because we had such a long runway. Adding to the pressure, the success or failure of a project within drug companies often hinges on the dedication of one person—an individual who is willing to stick his or her neck out and champion a drug when no one else believes in it.

Oddly enough, pharma corporations are aware of this. Back when I worked at Pfizer, an internal book was written about every major drug that had been discovered at the company. In each case, there was a point in time during development when it looked as though the drug would be killed—maybe even should have been killed based on the data. But none of the projects were stopped. For every successful drug, there was an internal champion who pushed to keep the project alive. One of the examples includes the near-death of Lipitor, a drug that eventually went on to make the company around $13 billion per year and prevent millions of heart attacks. But it's hardly the only case of a determined scientist championing a drug in the face of organizational opposition. The documentary *Breakthrough* chronicles how the biology professor Jim Allison created the modern field of immuno-oncology through sheer will by convincing skeptical pharma companies, the FDA, and investors to advance his drug. Despite the personal losses he suffered, the result has been millions of lives saved and a Nobel Prize.

As a scientist, it can be disheartening to hear these stories and imagine all the drugs that could have been. How many Lipitor-like projects and cures for cancer have been killed because they lacked a champion?

So shortly after securing our office space, we began searching for scientists who could champion our discovery efforts. One of the most important hires we had to make was our Chief Scientific Officer (CSO). The CSO plays a critical role in any biotech company, overseeing and managing research operations and ensuring that projects are staying on track. Generally, a person in this position has a PhD—and either a wealth of experience in the industry or a former career at a world-renowned research university.

As with most processes at Morphic, we approached hiring from a non-traditional angle. While the majority of CSOs tend to be biologists, our platform was more chemistry-focused and bolstered by the strong biology knowledge of our founder. Our CEO Praveen—being the type who keeps an eye out for overlooked opportunities—thought that since most CSOs are biologists, there must be a large, untapped pool of qualified chemists out there that would be better for the role. We decided to use that to our advantage and began hunting for a chemist in August of 2015.

Since I was the Head of Chemistry, and the most senior scientist at the company, Praveen made me responsible for vetting the candidates. The search was a bit nerve-wracking, though, because I was hiring my own boss. (It's tough to complain about a terrible manager if it's someone you've personally selected).

My biggest fear when it came to hiring a CSO with a chemistry background was that our experiences might clash. I'd spoken with other scientists in my position who had worked with a chemist CSO before, and almost all the stories ended badly. The common theme was how quickly tension could build between the two roles, since competing visions for the correct approach created conflict. High friction could lead to catastrophe.

Despite these risks, the rewards of untapped potential and the right expertise to advance our platform was much more promising. In order to find the right person for the role, we reached out to a headhunter named Josh Albert at the firm Klein Hersh. Josh had originally placed me at Cubist, and while he isn't a chemist himself, you wouldn't know it based on his knowledge of the field. In fact, he seems to know everyone in chemistry. Together, we assembled an internal hiring group that consisted of myself, Praveen, Bob, and one of our scientific advisors, Mark Murcko—the same person who had connected me to Kevin and, ultimately, to Morphic.

Hiring our CSO was a complex, drawn-out process. While it's easy for candidates to list accolade after accolade on paper, biotech is a small world. Often, one of us would veto a candidate based on what we knew or had heard about them from a trusted source. Once we asked Mark about a candidate, and heard him say: "No...No...No." I was thankful, he had fired one of my friends. Other times, a person we truly wanted to hire would be wooed away by another company.

One evening, during the height of our search, Praveen, Bob, and I had dinner with an ex-professor we were considering for the role. In many ways, he embodied the stereotypical professor persona and ego I was wary of. He was very matter of fact. Aggressive, even. I remember him saying, "In a small company, there's no time for career development. You need people who can do the job, and you need them to do it as fast as possible." He also said that when people ask about career development, he says: "You are privileged that you have a job; get to work!"

That was truly the opposite philosophy Praveen was trying to instill at the time, and I remember thinking, "I bet this guy is going to be very successful in biotech, and I'm pretty sure I don't want to work for him." It ended up being a moot point. He took a CEO role at an up-and-coming biotech startup that went on to sell itself twice for over $300 million (each time!).

Months later, as our efforts began to seem futile, Josh told us about a chemist named Bruce Rogers. I vaguely knew Bruce from my time at Pfizer, but we hadn't considered him thus far. When I worked there, Bruce had led a chemistry group and didn't have enough of the necessary experience to step into the CSO role. Yet after I left the company, he had moved into a venture group within Pfizer and expanded his responsibilities and leadership capabilities to include exposure to small companies, biology, and medicine.

It was my job to conduct the phone screening for new candidates, and if it went well, I would recommend bringing them on site. So, I hopped on a call with Bruce within a few days of Josh's suggestion. All I remember is that Bruce interviewed extremely well. Although he didn't have experience at a startup company, he quickly figured out what mattered to us. So at the end of his interview, he spent some time describing his work with smaller companies in his venture role at Pfizer.

It was true he hadn't been a CSO before, and he wasn't as experienced as some of the other candidates. But our initial conversation went so well that I couldn't help but be excited. In fact, I didn't realize

I was in such high spirits until our weekly hiring group meeting. I kept talking about Bruce and his potential throughout the meeting, and finally, Praveen stopped me to say, "You seem really enthusiastic about this guy."

I paused to think about it and realized he was right. I *was* excited about bringing him on. Our shared experience at Pfizer also meant we "spoke the same language." I knew we'd have a similar perspective on what was important and what wasn't in running the science, and I hoped we wouldn't fall victim to the traditional tension between a chemist CSO and Head of Chemistry. He may not have checked every box on the list we initially drew up, but it was clear he was the right fit.

Five months after the search began, we had our CSO.

CHAPTER 5

Have Biotech Startup, Will Steal Scientists

Landing the right CSO was a good start, but Bruce wouldn't be the chief of much unless we could hire scientists to lead our projects and design new drugs.

Without a strong chemistry team that could create molecules that held the potential to become drugs, none of what we were doing would matter. We were looking for chemists who were skilled at making molecules and who also had prior success in leading projects to a clinical candidate. While neither task is easy, the majority of chemists are taught how to make molecules in graduate school. Finding someone with experience leading projects is much more difficult. A company has to trust a chemist enough to place them in that role, and then the chemist has to be successful in that position. The problem is, once a good chemist has gained experience leading projects, he or she will often have plenty of interesting career opportunities to choose from.

Aside from experience, the size of our company was another complicating factor. It wasn't easy to convince people to leave comfortable positions at large pharma companies to come join a biotech startup. Building the team was going to be a laborious yet delicate process. So we had to persuade people to come on board early, knowing that the first few hires can make or break a small company.

On our original team was Kris Hahn, who was a superstar at Cubist. Kris was instrumental in the day-to-day logistics of setting up our lab and getting the ball rolling. It had been years since I'd worked in a lab on a regular basis, so it was essential to have someone on the team who could get the right equipment, obtain safety approvals, manage supplies, and do the 1,001 tasks that are involved in setting up a world-class chemistry lab.

Also essential was the first PhD chemist we brought on, Dawn Troast. She was introduced to us by her husband, who was in my group at Cubist, and she was a key early scientist. Dawn believed in us enough to take the job while the company was still largely unproven. She added firepower to our chemistry team from the start, has been an important contributor throughout the company's existence, and has played a key role in advancing one of our first clinical candidates.

There were several more major hires during the early phase of preparation and expansion, and none more impactful than Alex Lugovskoy. We were originally considering Alex for our Head of Biology, yet he was hesitant to accept the position. He was more interested in a development position, which typically comes during human trials—and we were years away from those. Even though we didn't have an obvious development position for Alex, a scientist at Biogen (a large Cambridge biotech where Alex had worked) told us he was the smartest person there. And he had actually worked on integrins previously, which was a rarity.

So Praveen made the early decision to hire this incredibly bright guy as our Chief Development Officer and figure out exactly where he would fit in later (Alex actually started just before Bruce). We soon learned the Biogen scientist wasn't lying. Alex is wicked smart, with a PhD in biophysics from Harvard and a wealth of knowledge in a number of fields within pharma unrelated to his degree. He's extremely hard-working and highly focused, but is still quite friendly and strives to form genuine personal relationships—not the type of person you come across every day. He ended up being a major part of both our fundraising efforts and our strategic decisions when choosing which biological targets to work on, all while waiting for those clinical trials to emerge.

After filling these positions, we heard about two talented chemists named Bryce Harrison and Matt Bursavich who had recently entered the job market. Both had been working at Forum, a now-defunct biotech company primarily committed to finding cures for neurological diseases like Alzheimer's. The company was unusual because it was backed almost exclusively by Fidelity, and in particular, its billionaire founder Ned Johnson, who had a passion for curing neurological diseases. Forum never had to scramble for investors, and its focus was clear from the inception. Yet after more than a decade of work, a failed schizophrenia trial was the final straw for the Johnson family. As so often happens in biotech, 130 people were suddenly looking for new jobs.

But one biotech's loss is often another's gain. When I called the head of chemistry at Forum to discuss Matt and Bryce, he told me: "They were the best two chemists at the company. If I were you, I'd hire them both."

The problem was, we didn't need both of them at the time. Yet being several months into the hiring process, we were well-acquainted with the realities of the biotech job market in Boston. When you have the opportunity to nab talented scientists, you take it. There's no guarantee the person you need will be available a month, or even a week, later. Matt, for example, had an offer to be the Head of Chemistry at a small virtual company. He ended up choosing Morphic partly on the advice of Al Robichaud, one of our SAB members and the CSO at Sage Therapeutics. Matt had worked with Al several years prior and asked him what he thought of the two companies. Al told him two things: Morphic had the potential to get a drug to market. The other company did not.

Luckily, hiring both Matt and Bryce worked out well, because we quickly realized we needed more capability to design the compounds we'd been dreaming up. However, as is often the case with biotechs, the design would happen in our own labs—but most of the production of those compounds would be handled externally by labs in China.

As a small company, you have to be more flexible in how you get things done. Many biotechs choose to work with chemistry teams in China for a number of reasons. For one, the Chinese companies employ very capable lab-based chemists who are great at making compounds. They're also extremely productive and cost-efficient relative to the companies a biotech could outsource to in America. And because of the nature of the work and the analytical data available, it's fairly easy to tell if a company has made a correct, pure compound.

However, none of those benefits will matter if you don't choose the right company and get a solid team to work on your compounds. The first Chinese company we worked with was ChemPartner. The company had been a small investor in Morphic early on, so they had a strong incentive to build a good team for us to work with. I had also collaborated with them back when I was at Cubist, so I felt comfortable working with them.

The partnership worked out well. Initially, they were only making compounds for us, but we eventually shifted our first biological assay to them as well. An interesting aspect of the Chinese companies is that they're not as strong in biology. They can repeat assays for you if you send them the exact procedure. But they're not as good at *creating* assays. This is where Terry Moy came in. He was one of our original scientists (an early pickup from Cubist), with a PhD in cell biology from Harvard. He created the first assays at Morphic and then transferred them to the Chinese scientists. That way, once they had created a compound, they could also immediately test it to determine potency. They would then email that data to us—all before the compound would arrive in the mail.

Eventually, we expanded our outsourcing to include more Chinese companies. One, BioDuro, we used initially for PK testing. The other, Wuxi, we partnered with for making large amounts of intermediates, and eventually for final compounds. Bruce knew their Head of Chemistry well, and they've been an excellent partner so far. Using multiple companies allows us to balance their strengths, creates some competition, and gives us increased stability should one fail.

Our experience working with the Chinese companies has been positive since day one, and much of that came down to the way we manage those relationships. The biggest worry for any company outsourcing work is whether or not the team working on the project will be truly engaged. The difference between true engagement and a lack thereof really depends on the strength of your relationships. It certainly helped that Bruce and I had connections to these companies before we started working with them, and many of us go visit the labs twice a year in order to meet and talk with people face-to-face. Plus, giving them Morphic T-shirts, and awards to their best scientists.

There's a very specific reason for that. As a small company, one of the keys to getting the best possible work from your partners is to make sure they know your company is growing and could become an important client later on. You also have to be willing to tell them to remove chemists from the team if something isn't working—even going so far as switch partners if you have to.

We were fortunate that our partnerships worked out so well. It was one less thing to think about as the company began to take shape and our focus on creating superior compounds intensified.

Chapter 6

Hitting The Target

In biotech, drug design is the process of crafting a promising molecule on paper in order to physically build it later. That means figuring out the exact connections of the carbon, hydrogen, oxygen and nitrogen atoms. Think of it like creating the perfect key for a very complex lock. If the key's too big, it won't fit. Too small, and it will wobble around inside the lock without doing anything. Even when you think you have the perfect key, you might find out that the body metabolizes it too quickly or you hit a target you weren't intending to, and the molecule is toxic.

Making molecules is an incredibly laborious process. Some of them can take months to create, and there's no guarantee they'll work the way you want them to.

Fortunately, we had something of a secret weapon—a partnership with Schrödinger and access to their state-of-the-art 3D modeling software. Schrödinger was one of our original partners, investing in our seed round and trading their services for equity in Morphic.

Today, roughly half the team at Schrödinger is made up of hardcore software engineers while the other half comes from a drug discovery background. The company is so effective because of that blend of expertise. Their ability to model compounds and predict their efficacy based on the crystal structures Albert and his group have created has been critical to our success. The Schrödinger team takes the crystal structures, and by using massive computer clusters they're able to analyze how the integrin will move, what new compounds will fit into it, and, ultimately, whether or not we've created the right key for the lock.

They call the process FEP, or free energy perturbation. Roughly, it's a physics-based algorithm that creates a precise conformation of a compound and the protein it needs to "dock" with. Schrödinger's software is a step above any other technology because this process can actually predict the potency of a compound in many cases. It takes an enormous amount of computing power, but it leads to models that allow both the protein and the drug to move as they would in reality. The models aren't perfect by any means, but they're effective. Their predictions around potency give us much more certainty about which compounds to make and which to skip. In essence, we're able to get the right compound with the right potency much faster.

It's been a hugely successful collaboration, which really depends on transparency and open communication. For instance, both sides are working to come up with ideas for compounds together. If one seems promising, Schrödinger will model it and then tell us whether or not it modeled well. If it looks good, we get to work making the actual compound. Sometimes, however, we'll make a compound even if it doesn't model well, in order to test a certain hypothesis. Other times, Schrödinger will ask us to make compound just to test their modeling software and prove that it's working. The collaboration works so well because each side gets something of value. We have an advantage when it comes to finding the best compounds, and the experience improves their systems and helps them create new software tools.

While our partnership with Schrödinger is important today, it was essential in the early stages of the company when we were trying to find our first leads and get started with minimal resourcing. Our first project was the multiple sclerosis target, $\alpha_4\beta_1$. Shortly thereafter, in December of 2015, we began work on $\alpha_4\beta_7$, our key inflammatory bowel disease (IBD) project. Around the same time, we made a handful of compounds that led to our initial α_v inhibitors that would be used for fibrosis targets, $\alpha_v\beta_1$ and, later, $\alpha_v\beta_6$. Looking back, this was a hugely productive time at the company, and laid the molecular foundation for everything that would come next.

Our future IBD drug would work by preventing the body's own immune cells from mistakenly attacking the healthy cells in the intestine. For those who've never known someone with ulcerative colitis or

Crohn's Disease, the implications for creating a daily oral drug to treat IBD may seem abstract. But for the estimated 1.6 million people who live with IBD in the U.S. alone, an oral drug would be life-changing. While a number of FDA approved drugs to treat IBD are currently on the market, many are biologics—drugs that must be taken intravenously every few weeks. Praveen often compares taking intravenous drugs to going to the dentist. The visit is generally awkward and uncomfortable, and most people find that monthly appointments test the limits of their endurance because it requires a lengthy trip to a medical facility, and reminds you that you are sick. It's like going on an extended visit to the dentist every month for the rest of your life. Obviously, the ability to wash down a pill with a glass of orange juice every morning would be genuinely life-changing.

To make matters worse, about a third of patients don't initially respond to biologic IBD therapies and another third will lose effectiveness within the first year. Clearly, more progress is needed.

In the fibrosis realm, we were aiming for a similar solution by initially pursuing $\alpha_v\beta_1$—a target thought to be useful for treating diseases characterized by excessive tissue fibrosis, such as liver fibrosis and Nonalcoholic Steatohepatitis (NASH). Closely related was $\alpha_v\beta_6$, where our compounds always seemed to work better, and so it soon took the lead. This integrin is most associated with lung fibrosis, in particular Idiopathic Pulmonary Fibrosis (IPF), which often takes life within 5 years of diagnosis, and there are no truly effective therapies.

These exciting new projects aimed at mitigating the formation of fibrosis, which is like scar tissue. To understand its impact, think about the last time you had a deep cut on your skin. When the body is wounded, $\alpha_v\beta_6$ is expressed, which sets off the deposition of collagen—a protein that acts as the underlying structure that allows your body to heal. You may have some aesthetic issues with the resulting scar, but you'll probably be fine. Yet when that same process repeatedly occurs in an organ, it can become problematic. Collagen and other proteins can build up on the organ, resulting in fibrosis and decreasing the organ's ability to function. So $\alpha_v\beta_6$ is crucial to wound healing, but it also needs to be checked when it's no longer needed. So our $\alpha_v\beta_6$ project is aimed at creating an integrin-inhibiting drug that can reduce or stop fibrosis in patients. A successful therapy here could help manage life-threatening fibrosis diseases that affect the lungs, liver, and kidneys.

Eventually, that project became the basis for our collaboration with a multibillion-dollar pharmaceutical company. However, as 2015 came to a close, that deal was still over two years away. Nothing was certain yet—and we had years of design work ahead of us.

CHAPTER 7

No IP, No Projects, No Money

By Q1 of 2016, we had nearly doubled our team and established our culture. Yet our Series A round of funding was still months away from being a done deal. It was a transitory period of time when we had to bear down and really focus on advancing our nascent projects down the long road to a drug.

One of the main issues during this period was a conflicting view of the next important step we needed to take. Our lead investor told us the number one thing we could do to give other investors confidence: prove our integrin conformation theory (bent is better than open) *in vivo*. That is, we needed to show it worked in animals. *In vivo* testing is often useful for researchers as a tool to test the efficacy and toxicity of a molecule. Investors would prefer it happen early in the process because they like to receive the news that a target doesn't work *before* making a hefty financial commitment.

The problem was that none of us were very confident in our ability to provide neat and tidy answers from the *in vivo* tests. Assays in animals are notorious for having large error bars, our molecules weren't very good yet, and we feared that the result would be uninformative or even misleading. Poor results at that stage could have killed the company or, at the very least, hurt our prospects and delayed advancement by months. Bruce, Alex and I knew all of this, and none of us were keen on diverting our efforts to what would essentially be a basic biology question. Our opinion was that we should put our efforts into areas where they'd be most useful—advancing projects toward drugs and building maximum value with our limited resources. Then, work to prove our hypothesis once we had better molecules for the diseases we were trying to cure.

In fact, that wasn't the only time we'd have to make difficult decisions on how to continue advancing toward our goals. Our original MS project, $\alpha_4\beta_1$, was crucial because it was how we achieved proof of concept for the entire integrin platform. Yet by March of 2016, it was becoming clear that $\alpha_4\beta_1$ was not a target that would get us closer to a drug. Praveen made the decision to kill the project—effectively ending our most advanced project at the time—and focus on $\alpha_4\beta_7$, our IBD target.

It was a tough choice, but the rationale was clear. Hitting $\alpha_4\beta_1$ was going to have a huge risk of causing the PML issue that had sunk Tysabri and Raptiva years earlier. Although the target was validated to work in MS, a raft of new approvals in that area made the space very competitive, especially if your drug had a bad side effect. Our IBD project also had one extremely valuable advantage that made it a compelling target from both the business and scientific lens: it had already been biologically validated by an IV infused drug. Which meant we should know after Phase 1 trials whether or not it could be an oral drug. This would be the early value inflection point investors would love, and a chief selling point for the company.

Still, nothing is ever a sure thing in the biotech business. While we had a strong selling point, we hadn't filed any patents yet, which are also called intellectual property (IP). Filing patents starts a 20-year clock on how long a company can sell a drug, and it also lets competitors know what you're doing. We had made the conscious choice to stay stealthy and maximize the value of our drugs once we found them. However, investors often want to see that a company has patent protection on its technology because it validates novel inventions. So we found ourselves at a point where, as Praveen liked to say, we had no IP, no projects, and no money—at least until we could close our Series A round.

Fortunately, we were able to convince our investors that in vivo testing would just be a distraction at this stage of the game, and that we now had the right targets and strategy. Amir Nashat, one of our founding board members from Polaris, deserves credit for bringing the rest of the board along. And the Series A went just fine without those tests. We wound up securing funding from all our original investors and several new ones: Pfizer, AbbVie, Omega Funds, and SR One—the venture arm of GlaxoSmithKline. The Managing Director we worked with at Omega Funds was a flamboyant Italian by the name of Otello Stampacchia, who was very particular about espresso and even brought his own machine to board meetings. Though we didn't know it at the time, our relationship with Otello and Omega Funds would be crucial to our Series B.

A contagious sense of excitement was buzzing around Morphic while the Series A was in progress, but closing the round was a major relief. The new investors extended our runway and gave us the opportunity to expand our team and hire the people we needed.

The catch was that our Series A had three tranches, meaning we only received about one-third of the money upfront. Tranche investing is a common method used for biotech Series A rounds since investors don't quite trust a drug discovery company that hasn't proven itself. They're invested, but they also don't want to lose all their money if the company doesn't perform or hit deadlines. If they don't see progress, investors can back out and only lose a portion of their money. Because of that option to pull out, this type of investment can easily push a company under if it doesn't deliver and can't find another source of funding.

In order for our team to reach its next tranche, we had to advance a project to lead optimization, essentially the point where you have animal data and can predict a human clinical candidate will be made in 12 to 18 months. After killing the $\alpha_4\beta_1$ MS project and deprioritizing $\alpha_v\beta_1$, we had firmly placed our hope in $\alpha_4\beta_7$ (IBD) and $\alpha_v\beta_6$ (fibrosis). With a short timeline to running out of money, we were also highly motivated.

Unfortunately, we were still missing one critical piece to getting there—a Head of Biology. Since we were initially building a structure-focused company and had opted for a chemist CSO, we simply didn't have as many connections in the biology world. While we worked with an executive recruiter to help find candidates, the search wasn't easy. It dragged on for so long that Bruce began hiring the biology group without the Head of Biology in place. That's generally the responsibility of the biologist, but we couldn't wait any longer to get started while we looked for the right person to fill the role.

The search came down to our eventual hire, Liangsu Wang, and another candidate we also liked. Tim wanted to meet them both because it was such an essential hire, and he came away from the meetings convinced Liangsu was the woman for the job.

Since we had begun preemptively hiring the biology group, Liangsu wound up walking into a culture and team she didn't have a hand in creating. The group had been reporting to Bruce up until that point, and they'd had a bit of leeway when it came to their supervision. Liangsu didn't waste any time. She made sure everyone was focused on the right aspects of the biology and working diligently to move projects forward.

The hard work paid off. By July 2017, both $\alpha_4\beta_7$ and $\alpha_v\beta_6$ had reached lead optimization, opening up our second tranche of funding. Both were in an excellent position, but since $\alpha_v\beta_6$ was slightly further along, it became our first shot at a development candidate.

The work ethic and thoroughness Liangsu brought to the team was a crucial element to the culture we were laboring over, creating piece by piece. To some extent, everyone in the company was involved in creating the culture, but company leaders always play a critical role. That's why we only chose people who lead by example and were focused on building what we called a "truth-seeking" culture.

CHAPTER 8

Out Of The Comfort Zone,
Into The Culture Shock

In science, the easiest person to fool is always yourself. The path of least resistance is to disregard evidence that doesn't support your desired outcome, especially when there are millions of dollars riding on the result. You see it happen at an organizational level all the time at big companies. Everyone knows the company wants to move in a certain direction, and it doesn't take long before scientists, executives, and hiring managers begin falling in line—even if the evidence is pointing in a different direction.

From the start, our goal at Morphic was to create a culture where people are comfortable seeking the truth, giving feedback, and voicing their honest opinions. In fact, Praveen told us we could disagree with him at any time, even in a board meeting! He encouraged everyone, at every level, to speak up and say what they're thinking and feeling.

But for that type of communication to thrive, the culture has to revolve around transparency and communication. So at Morphic's inception, Praveen met with Albert and I to make two stunning announcements: no email, and no PowerPoint. I was shocked. Instead of email, we would communicate via the messaging platform, Slack. And instead of PowerPoint, we would all write out our presentations in a formal document, using full sentences!

The choice to use Slack, while a little out of the ordinary, was actually quite reasonable. Email has a tendency to get out of hand, and most of us had plenty of experience with bloated inboxes full of irrelevant emails. Also, the question of who to 'cc,' or who you have *forgotten* to 'cc,' has destroyed many relationships. However, Slack was a colorful, playful messaging system that felt inherently casual while spilling its contents for all employees to see—or not—based on their personal channel settings. Slack certainly took some getting used to, but people quickly realized how good it felt to be freed from the tyranny of the inbox. As an added bonus, the group chats and instant messaging vastly improve transparency across the company.

PowerPoint was different. When I heard we wouldn't be using it, I was incredulous. I'd spent the majority of my career using slide shows to summarize practically everything I needed to present. At the previous companies I'd worked for, PowerPoint had been an important tool when it came to defending my work or trying to advance a compound or project. But Praveen viewed it as a crutch—something people used to simplify an argument and allow the audience to see without really engaging. He argued that writing ideas out in full sentences and paragraphs forces active participation on the part of both the presenter and the audience. You have to make a convincing argument through prose, and the people in the room actually have to read it and think about it critically. Meeting time is then spent on discussion, questions, and decision making. So, PowerPoint was nixed.

When it came to advancement, the common system of performance reviews was out, too. Truthfully, I had always been dissatisfied with the way performance management had been handled at my previous companies, but I thought of it as a necessary evil. While we still hold annual reviews at Morphic, we also have quick quarterly reviews as a way to keep the evaluation focused, balanced and relevant. When it comes to evaluating the team as a whole, the people who had a blowout year—maybe the top 10%—get extra bonuses. But we're not spending time figuring out exactly where everyone falls. If you're doing really well, you'll be rewarded extra. If your performance is consistently poor, you may be let go. At a small company, it is quite clear we are all in the same boat, and will succeed *or* fail together. Obviously, this is a huge performance incentive in itself.

Nixing traditional processes weren't Praveen's only ideas for building a great culture—not by a long shot. For instance, he likes to take notes on a waiter's pad, eschewing the Moleskine notebook or briefcase a typical executive wields. It's his way of reminding himself not to put on a show or worry about what other people think of him, along with showcasing some amount of civil disobedience. Naturally, he hands out waiter pads to every new employee.

He also instituted a program for career development that was available to everyone in the company. The gist of it was that each person on the team had to choose the job they might want to have in five years. Then, Praveen and other members of the leadership team helped us connect with someone who already had that job. Through this method, I spoke with Chief Scientific Officers who started out as chemists, picking their brains and trying to understand how they made the leap from Head of Chemistry to CSO. I also met with venture capital partners to learn how they moved from scientist to investor. It was really an effective way to learn more about my own potential career paths and better understand what was needed to get ahead.

The most important aspect of all of these idiosyncratic cultural decisions is that they were made at the very beginning. Praveen actually had a list of 20 problems that all companies ran into eventually. He wanted to get ahead of them while Morphic was a clean slate, rather than later on when we'd have to break ingrained habits. For instance, if they had instituted Slack when I was at Cubist, people would have used it for a few weeks, but eventually, everyone would have forgotten about it and reverted to the email system. People were just too set in their ways by then. The goal at Morphic was to take these concepts and make them a part of the company from the very start. Very few emails or PowerPoint, no bad habits.

Building our culture was an ongoing project. We wanted to institute those cultural norms and values early on because we knew we would be too busy working on the science to go back and "fix" our culture. We had one shot at getting it right.

Fortunately, Praveen had a game plan.

PART III

Looking Ahead

Chapter 9

*Leading The Way to a Drug;
or The Race to Advance the Projects
before we Run out of Money*

To the outside observer, it may seem like our work at Morphic was happening in a vacuum, but there were (and still are) plenty of competitors racing to develop integrin-inhibiting drugs.

For the fibrosis target, $\alpha_v\beta_6$, the biotech company Pliant always seemed to be a bit ahead of us. We thought of them as our closest competition since they were founded around the same time as our team, though they had a head start in this target from their academic co-founders. However, we noticed one key factor that separated us from Pliant—they weren't working on a *selective* $\alpha_v\beta_6$, and we believed they were targeting the wrong conformation. As far as we could tell, they had a dual $\alpha_v\beta_1/\alpha_v\beta_6$ inhibitor, binding in a classical, opening conformation. Based on Tim's work, we believed that the opening conformation was inferior because it risked having the opposite effect—it could actually make the disease worse.

We saw a similar competitive landscape for the IBD project, $\alpha_4\beta_7$. We caught wind of several competitors working on the target, but their drugs were what we call "gut topical." You take them orally, yet hardly any of the drug is absorbed into your body. The way we understood the mechanism, it didn't appear that those drugs would have efficacy in humans.

The thing is, it's very difficult to get selectivity for the $\alpha_4\beta_7$ integrin, so historically many companies worked on a dual inhibitor that targeted $\alpha_4\beta_7$ and $\alpha_4\beta_1$. The problem with inhibiting $\alpha_4\beta_1$ is that it targets immune cell surveillance of brain and CNS tissue, instead of just the intestinal tissue, which can lead to PML—the deadly disease that initially shelved Tysabri and Raptiva. However, by selectively inhibiting just $\alpha_4\beta_7$, a drug will only affect immune cells targeting the intestinal tissue.

Achieving selectivity is no simple task. Half of the binding site is exactly the same, and the other half is so similar that it takes incredibly targeted compounds to selectively bind to the correct beta binding site. We didn't have time to worry about the competition too much, however, because we had plenty of progress to make in our own labs. We worked for a long time on $\alpha_4\beta_7$, examined several different drug scaffolds, and used every advanced computational and structural technology we could think of. It was really tough, but gradually Dawn and Matt started to make progress and crack this puzzle.

Although we were still exploring a number of projects, hitting lead optimization for both our fibrosis and IBD projects would be crucial, as it would open up another tranche of funding from our Series A investors. The tranche was tied to the lead optimization milestone because reaching it essentially meant that we had compounds we believed could become development candidates within a year or two. If we missed, we would run out of money by the end of the year. Investors needed tangible compounds to hang their hopes, dreams, and dollars on.

Ironically, the fibrosis $\alpha_v\beta_6$ project went a lot quicker. Bryce led this one, and it always seemed like he drew a dream hand. It wasn't easy, but his drug discovery skills led predictably to a good pace of successes and progress.

Both our investors and our scientists experienced a growing sense of excitement once we reached lead optimization at our board meeting in mid-2017. The truth is, we didn't really have any hard data to point to when we were putting together our Series A funding one year earlier. Investors were buying in to our scientific team, our founder, our CEO, our story. With lead optimization, that air of speculation was all in the past. We were producing solid data and had truly exciting molecules on our hands.

While everything seemed to be moving in a positive direction, in our line of work, success doesn't mean you can let your guard down. In fact, one of the major dangers in biotech (or any business, for that matter) is loss of focus. Almost all successful companies go through a period where they think that their technology can be applied more broadly than it really can. Once a company finds a taste of success and realizes its technology works, the initial reaction is to expand its work into new areas. The team starts taking on new projects and agreeing to exciting opportunities. Of course, success is never quite so simple. The company then gets diluted—spread too thin to make significant leaps in one area.

Eventually, the company has to take a step back and focus on what was previously working for them. That is, if they have enough funding to survive the setback. A poor sense of direction at that most critical of junctures can swiftly kill a company.

At Morphic, our board gave us an early warning about the side effects of dilution. While we didn't give up every project to focus on the fibrosis and IBD targets, we knew we had to make our focus clear. So when we rolled out our corporate goals in early 2018, the only two projects we mentioned were $\alpha_v\beta_6$ and $\alpha_4\beta_7$—the targets that had reached lead optimization. We wanted to make it clear to everyone that if there was any choice to be made, it would be decided in favor of advancing those two projects. Whatever happened, we were going to dance with the projects that brought us to the party.

Even with the thrill of the achievement buzzing through our labs, reaching lead optimization was by no means a time to relax. As the name suggests, it still requires a huge amount of work to optimize the compounds into development candidates.

For any compound moving from lead optimization to development candidate, there are at least 10 variables to optimize—but the three main concerns initially are its potency against the target, will it be orally absorbed, and the speed at which it's metabolized. Essentially, how well will this drug work, and will it get to and remain in the body *long enough* for it to be effective? To help us answer those questions, we brought on Dan Cui as our Head of Pharmacokinetics (PK).

Dan's job was twofold. First, he had to design the experiments that would help us understand what the body would do to the drug. He then had to help us interpret the data generated by those experiments. We needed to know what to change in order to obtain the most effective compounds. What would help the drug become more permeable and metabolize slower?

Of all our projects, the PK for our $\alpha_v\beta_6$ target was always surprising. We ran into an issue with the in vivo testing at one point because the compound was absorbed well in dogs, but not in any other species we tested. We were left to decide the scientific, and perhaps philosophical, question—is a human more like a dog or a rat?

As we continued optimizing both compounds, it was clear $\alpha_v\beta_6$ was progressing much faster toward becoming a development candidate. Advancing compounds isn't a strictly linear progression. Sometimes you improve one aspect of the drug and it's absorbed—but it isn't metabolized well. Or the opposite might be true. These issues tend to bounce back and forth, but, overall, we were consistently moving in the right direction.

As is usual in the biotech sphere, we weren't the only ones who took note—something that would become abundantly clear during the industry's largest annual investment conference.

Chapter 10

Let's Make A Deal

The science is always the primary focus of a biotech company. A company without a promising compound or two won't be around for long. But while the science is always front of mind, funding is never far behind.

I was never all that worried about raising our Series B round, to tell the truth. We had good data, we had committed investors, and it seemed as though all of our Series A participants wanted to be in on the next round—which they ultimately were. But the round took longer than expected to put together. The problem was finding someone to lead it. We initially thought a company based in China would do it, but they ended up backing out at the last minute.

Finally, Otello and his espresso fueled the funding. Omega Funds led the round with two new 'crossover' investors, Novo and EcoR1, and we were able to successfully close an $80 million Series B. Crossover is a term used for investors who anticipate a relatively near term IPO, which meant the pressure was on to go faster and faster towards that landmark event.

The time it took to secure the round was a little strange for us because, from our point of view, it was a no-brainer. With so many people interested in investing, it was unlikely that the round wouldn't come together. But we had to remind ourselves that nothing is ever 100% clear in biotech. Certainty goes out the window because there's always an element of the unknown—a chance (however small) that a deal might fall through.

Realistically though, we were in a strong position both scientifically and financially. And as our fibrosis and IBD projects continued to move forward, so did negotiations with multiple pharmaceutical companies that were expressing interest in partnering with us to develop our compounds. We had noticed this interest brewing in 2017 at the annual J.P. Morgan Healthcare Conference in San Francisco. But it wasn't until returning the following year that we began fielding serious inquiries.

AbbVie, one of our Series A investors and a pharma powerhouse, was one of the original interested parties, and they began to make significant overtures immediately after the 2018 J.P Morgan Conference. But by then, they weren't the only ones. Our compounds were looking better and better as the year carried on, and the competition rose to reflect the promising data.

Toward the end of the year, four prominent companies were kicking the tires on a partnership. The situation was unusual, to say the least. Many biotechs struggle to find even one interested party; suddenly, we had four knocking on our door!

I clearly remember how surreal the opportunity felt. It hit me when, around that time, I saw a former colleague from Cubist. He couldn't believe we had four bidders on the same asset and said, "I'm lucky if I can get one person to even look at us." It was true; we were lucky. But as businessman Eliyahu Goldratt famously said, "Good luck is when opportunity meets preparation."

While it seemed like an overwhelming opportunity for such a new startup, Praveen was in his element as CEO. The strong competition for a partnership gave him space to negotiate and push back against companies that absolutely overshadowed us in scale. And that was even *before* taking into account the quality of our compounds.

It's standard procedure for any company that wants to partner with a biotech to perform due diligence on the asset—the compound they're interested in. So the company generally hires a third party, a very senior chemist, to evaluate the asset and advise them on the potential strengths and weaknesses before partnering. It's a huge step in the negotiations, and fortunately, the chemistry consultants had nothing bad to say about our research platform, or the molecule taking center-stage. Bill DeVaul, a respected lawyer who was consulting for us at the time, said it was the first time he'd ever seen that happen. There was always *something* the chemistry consultants could ding and the partnering company could use as leverage. This time, they gave a clean bill of sale.

In the end, negotiations came down to the wire. We had two fairly evenly matched deals on the table for the fibrosis project, and it was certainly a difficult decision. Tensions flowed as companies battled to license the promising compound. Eventually, AbbVie won out in a three project deal, including $\alpha_v\beta_6$, and it was inked on October 18, 2018. Our three-year-old startup had just secured a $100 million upfront payment, and a potential total deal worth over $1 billion dollars, from one of the top pharma companies in the world!

Chapter 11

Clinical Trials:
Out Of The Frying Pan, Into The Fire

Despite the high of a successful Series B round and the partnership with AbbVie, we still had plenty of serious science to ahead of us. We put our heads to the ground, beakers in hand—and by the end of 2018, had created our first development candidate, MORF-720. This was quite a high point for the company. Bryce Harrison and his biology co-lead, Min Lu, made remarkable progress, and Jim Dowling, a senior chemist we stole from our landlord (AstraZeneca), gave us an extra push to get us over the finish line. All supported by a broader Morphic team of course.

The victory was sweet but short-lived. We quickly set our sights to our next goal: achieving Investigational New Drug (IND) status for MORF-720 by the end of 2019.

Once a compound reaches IND status, a company is allowed to transport the drug candidate across state lines and begin clinical trials in humans. The IND must then pass through the three phases of clinical trials generally outlined by the Food and Drug Administration:

- Phase 1 hones in on how patients metabolize the drug and what adverse effects it may have.
- Phase 2 begins to look at the effectiveness of the drug compared to placebos or other treatments, with researchers still keeping a close eye on any safety issues or side effects.
- Phase 3 focuses on gathering a lot more data about the drug's efficacy and safety, among diverse patient groups, and in conjunction with various drugs.

As you can see, the FDA is serious about patient safety, and many drugs don't end up working for the diseases they were supposed to. The rigor of clinical trials is viewed as a trial by fire, which is why a mere 14% of drug candidates wind up making it to market. Much of the work to even get a drug to IND status revolves around toxicology. Keep in mind that *everything* is toxic at some dosage, so the issue in development is really ensuring that the most effective dose of the drug isn't toxic in humans.

Different diseases also have different thresholds for the amount of harm a drug can cause while treating an illness. For example, treatments for cancer have more leeway when it comes to their toxicity because the patient is fighting a life-threatening disease. Patients with inflammatory bowel disease, on the other hand, are otherwise fairly healthy and may live for decades. The drugs used to treat IBD can't cause the type of harm that might be acceptable for a cancer treatment.

Beyond toxicology, the other issue companies have to solve for at this stage is scale. Drug discovery is done on a small scale, milligrams to grams, so scientists can get away with a lot of nasty reagents and arduous purifications. Up until this point, our work was being done by a few dozen scientists at one site with glass flasks that might hold a liter or two. We were able to use all sorts of toxic and explosive reagents that you really don't want to use in an industrial plant inside massive steel tanks. On a large scale, that work has to be extensively optimized by another type of chemist, the process chemist.

The large-scale production had to be external, of course. The goal of process chemistry is to optimize the efficiency and safety of our synthetic process to make one compound. They don't have to worry about biology, PK or anything else, so they are deeply knowledgeable in synthetic organic chemistry. We worked extensively with an excellent process chemistry consultant, Stephane Ouellet, a French Canadian who used to work at Merck. He was recommended to me independently by three former colleagues, all of whom raved about him.

Historically, Merck was known as the absolute best in this field, even successfully beating famous chemistry professors to synthesize new natural products. We were lucky to get Stephane's advice, and he quickly steered us to a Merck production site that had just been divested by the company. It was in the UK, and the site was under new management but had retained all the Merck scientists. It was quite the advantage for a new biotech, to be able to tap into the scientific expertise and resources of one of the very best pharmaceutical companies in the world.

Access to this world-class expertise was especially advantageous early on, when we needed to be a lean startup, and had none of this knowledge in-house. Eventually, we hired Hanh Nguyen and Kristen Mulvihill to lead these crucial activities, but early on, you have to do a lot yourself, even if you are not quite qualified to do so. This is one thing I really love about a startup, you are forced to learn fast, network with others, and broaden your abilities, with everything on the line.

I remember visiting the plant where the production occurs, my eyes widening at the massive vessels—sometimes thousands of liters— that contained the reactions. Even the slightest mistake at that scale is catastrophic. For me, those huge tanks were a striking visual representation of just how far we had come in just three years. From glass flasks to industrial-scale production. From a handful of people working from home to a humming laboratory on a pharma giant's campus. And yet, we still had years of work ahead of us.

Chapter 12

Growing Pains

At several points in Morphic's history, it felt like we were hiring all the time—and late 2018 was one of those moments. Some days we were bringing in multiple candidates for interviews, and with a schedule like that, it can be difficult to stay focused on day-to-day work.

We began 2018 with about 20 people, but by the end of the year, our team was 45 employees strong. Of course, cultural changes are guaranteed with that type of growth. The increase in staff and capabilities was positive, but it did change office dynamics in subtle ways. I walked into the office one day, and for the first time, I didn't recognize everyone. I can't always call up someone's name on the spot anymore.

The additional personnel also required additional room. Another biotech in our building had recently gone out of business, and we were able to snap up their space. As a result, the entire team wasn't in the same hallway anymore. Biology was in one building, chemistry and structural biology in another, leadership increasingly in a third building. Of course, the team began to worry about developing multiple distinct cultures. What happens when three groups don't interact as much as they used to?

Growth is a natural consequence of success, and a company simply has to find ways to deal with the challenges it brings. Today, it can feel crowded in the office sometimes, but I always tell people that it's certainly better than the reverse. I've seen companies lay off dozens of people in one day—heck, I've been laid off. Being crowded is better than being sacked.

As we've grown, I've been accused of wanting things to remain like the way they were at the beginning of Morphic. To some extent, that's been true. Most people naturally dislike change and disruption, and it's normal to miss the connectivity and raw early excitement within the organization as it grows. Even Praveen, who embraced the shifts of scaling, used to have an office where he would see dozens of lab workers breezing by every day. Recently, his office was moved elsewhere—and it is more isolated. We all began to miss the daily encounters that a smaller company provides, especially because they were harder to replicate as the company grew and people spread out.

Yet, it was hard for me to view any of that in a negative light. The biotech business is fickle. We can never really know with certainty which projects will succeed or fail. Since no biotech is really in control of its destiny, when we began to hit our stride, the excitement outweighed any growing pains.

In reality, all biotechs that experience any success experience a similar phase: as they begin clinical trials, expenses start to mount. If poorly funded, the company can face a choice. The discovery team, the people who created the molecule being tested in clinical trials, can be thought of by investors as less important than it used to be. Development is sucking up more and more resources, so if money is short, leadership may be forced to cut costs. Eyeballs start to turn toward the discovery team.

The question soon becomes this: do you give in to the temptation and cut the discovery team? After all, you have what could be a successful drug. And the payoff for keeping the discovery team intact would only come much further down the road. With all those forces in play, it's not unusual for the discovery team to be diminished. In fact, that's what happened at Cubist a few years before I got there. As a result, they had no drug pipeline and had to start over from scratch.

Fortunately, Praveen said from the beginning that he wanted to put discovery on a pedestal at Morphic, essentially doing everything possible to avoid an outcome where the discovery team would be cut. So as our compounds moved forward and work began on development, we took a different approach. We have kept discovery intact. We knew our development group was only so big and could only handle so much work at one time. So while development worked to advance the current compounds, discovery could focus on creating new molecules that could potentially become drugs further down the line.

Of course, the money to keep funding discovery has to come from somewhere. The idea all along was to ensure steady funding for discovery to avoid this trap. The Abbvie collaboration was a three target deal, ensuring that two more projects would be funded, but that wasn't enough for the whole department. Fortunately, a new partnership was beginning to materialize.

Johnson & Johnson's pharmaceutical subsidiary, Janssen, had been a player during the negotiations for the fibrosis target, $\alpha_v\beta_6$—even though AbbVie was the eventual partner for that deal. Still, Janssen continued to express interest in working with us in the integrin space, and we did have several connections within the team. Liangsu's former boss is the head of the fibrosis therapeutic area at Janssen, and Praveen happened to know their CSO very well.

Janssen came to us with three integrins they were interested in—targets we had never worked on up to that point, which belonged to a different subclass of integrins. This partnership was exciting scientifically because we were stepping into this brand new area when it comes to basic science. We began targeting a different part of the integrin we had never worked with, which essentially meant starting from scratch. However, they're paying all the bills, so the long-term risk is much lower. If we prove to be successful, it will significantly expand our technology platform and open up multiple new opportunities in the future.

Discovery, it turns out, was worth putting on a pedestal after all.

Chapter 13

Potential For The Pedestal

Even with the heady proposition of two strategic partnerships on the table, we're still extremely focused on our promising—and absolutely critical—IBD project, $\alpha_4\beta_7$.

One of the great things about $\alpha_4\beta_7$ is that we actually know it works in humans. Many early-stage biology ideas look promising when they examine compounds in mice or test tubes, but they're often disappointing outside those parameters. In part, that's why it's so important to have the backing of outside partnerships. It's much easier to spend time working on a high-risk target when someone else is paying for it.

Beyond the partnerships and IBD project, there's also evidence that if we inhibit the right integrins, we should be able to stimulate the natural immune system to attack cancer. Immuno-oncology is very hot in cancer research at the moment. In layman's terms, immuno-oncology refers to helping the immune system realize cancer is foreign to the body and then coaxing it to attack and kill the cancer cells.

Researchers have been interested in the possibility of using the immune system to fight cancer for years, but a decade ago, the consensus was that the technique was probably a bad idea. A healthy immune system is already well-balanced and sits on a knife's edge—if it becomes too activated, a patient can get an autoimmune disease; if it's inhibited, the patient can end up with a serious infection.

It's interesting to see these ideas coming full-circle as people continue to work on them. And using our work to potentially treat cancer is a particularly tantalizing prospect for me, personally. There's a huge medical need there, millions of people are dying, and I've always been idealistic. Truthfully, the thought of being a part of the cure for cancer was a powerful motivational force in my career. My graduate work focused on a marine natural product that could be used in cancer treatment. Later, a company actually formed around that product, and it continues to this day. They have a lot of interesting data, but the work is incredibly complicated.

Oncology was also the first therapeutic area I was a part of when I got to Pfizer. I worked on many projects with varying degrees of success. The first project I led was called AKT. At the time it was an extremely promising target for cancer, and we had a lot of success and achieved a development candidate. Ultimately though the mechanism failed in the clinic, due to toxicity and a powerful feedback loop in humans that resists inhibition of that pathway. That is often how it goes—hard work, great scientific insight, thrilling success, then failure. Often for reasons that couldn't be predicted earlier.

However, at the end of my time in the Pfizer oncology group I had one great success. A biology colleague (Gary Borzillo) and I started a project that actually turned into a drug. I led the early project, but as so often happens, I was moved out of the therapeutic area during one of the regular reorganizations within the company, and I had to hand off the project to other scientists. They did spectacular work, and the drug, Daurismo, was approved by the FDA in 2018 for leukemia.

The new therapeutic area I landed in was anti-bacterials, which is probably what got me the director job at Cubist, which is where I met Praveen. And that path eventually led to my position at Morphic, where we're now looking at the potential for integrins in immuno-oncology. The grand circle of biotech life.

Chapter 14

Getting Stock In The Science

Our partnerships with Janssen and AbbVie were just that—partnerships with both sides collaborating to make the most of the opportunity. However, that's not always how it works when a larger company is interested in a startup biotech's molecules.

Morphic was originally set up as an LLC (limited liability company), with each project—like $\alpha_v\beta_6$, for example—acting as its own independent company. The advantage to this approach is that if a larger company is interested in acquiring a startup's project, they can buy only that project. The team working on it sometimes goes along with the purchase, if the acquiring company still needs the productivity and knowledge-base of the project team. The caveat here is that the acquiring company doesn't have to buy the entirety of the startup, including projects that they don't want, so the team could continue normal operations to advance its other projects. So, nobody gets fired. In fact, the startup often gets enough money from the acquisition to finance its other projects for a couple of years.

An example of this practice in the wild was when Gilead—a multi-billion dollar pharmaceutical company—bought a NASH drug from Nimbus, a smaller biotech, for $400 million upfront. Investors and employees shared in the reward, no one got fired, and work continued on other projects with more resources.

From an accounting perspective, this setup isn't necessarily ideal. The finance team has to continually deal with the tax implications—not just for the company, but for all the "sub-companies" that each project entailed. Paperwork aside, an LLC will not work for a company that wants to hold an IPO (initial public offering). So once we began to consider an IPO near the end of 2018, Bob and the finance team started folding all the sub-companies into one, creating what's known as a C Corp.

While we knew that an IPO was an eventual possibility, the catalyst was really our Series B round when we brought in what are known as "crossover" investors. In earlier rounds, our investors were focusing on the long view and were prepared to wait longer for a potential large return on their investment. Crossover investors, on the other hand, are people looking for a return in a shorter time frame—usually within a couple of years. The best crossover investors are also prepared to invest in and support an IPO at a later time. That was important if we ever went public, because there needs to be a significant demand for shares in order to keep the stock price from tanking.

Even with the Series B closed and a solid lineup of investors backing us, going public wasn't as easy as having Bob file a few tax forms. One of the requirements for a company holding an IPO is that they file an S-1 form with the SEC roughly a month before the IPO. The S-1 includes all sorts of details about the company—non-confidential research, previous financing, and, interestingly, the compensation of everyone in the C-suite. So Praveen, Alex, and Bruce all had to publicly disclose their salary, bonuses, shares in the company—everything. To tell the truth, I was more than a little relieved to not be put under the microscope. Not that many were spending time digging through all those pages, but it isn't something most people are thrilled to have completely out in the open.

In general, numbers take on new significance both in the lead up to and the aftermath of an IPO. When you own shares of a company that's going public, it's difficult not to keep one eye on the numbers. Employees are locked out of trading for six months after the IPO, yet many continually track the price.

The initial price for Morphic's stock was $15 a share, though the first trade was actually at $18. Based on the strong demand, we upsized the number of shares sold and raised nearly $100 million dollars. Our market cap is $470 million as of writing this, but there were times when it was close to $1 billion. Those massive swings in valuation keep everyone on their toes, but to the management team's credit, the stock price has never been a key focus of the company. The intent is still to keep everyone focused on the day-to-day work because our dedication to the science is what got us here in the first place. The goal is (and always has been) to help patients. If management were to begin focusing heavily on the stock price, that could lead to a change in the atmosphere. Once it seems like the company's goal is to make millions and cash out, the entire culture, and the future of the company, is in jeopardy.

The other reason you can't spend too much time focusing on the stock is because, inevitably, you're going to have good data and bad data that will affect the price. Say we get good data for the $\alpha_4\beta_7$ project, and that is shared publicly. That news will send the stock price up. But if the results indicate some significant issue, the price will likely tank. Neither scenario is indicative of the larger future of the company, and we've had plenty of ups and downs before. But now, we have to publicly share anything that is financially material to the company. If the team gets too caught up in worrying about the stock price, normal setbacks or steps forward might begin to seem more critical than they really are.

Most people on our team had never been through an IPO before, so the organizational changes and the regulations surrounding it were new to them. That meant everyone had to be informed and educated about a number of the steps we were taking.

To get people up to speed in preparation for the IPO, the company executed a financial maneuver called a reverse stock split. This process condensed the current shares of the company into a smaller number of shares, each of which is worth more than it previously was. The reasoning behind the move was that we didn't want to have hundreds of millions of shares that were each valued at $2.00. By decreasing our number of shares, we were able to simultaneously increase their value.

After our reverse stock split, for every ~5 shares someone owned in the company, they would now own one share—though that share is still worth the same as the previous 5. If you don't tell people what's happening or why, their first reaction is going to be panic at the decrease in the number of shares they own. So, the management team had to spend time ensuring everyone understood that their overall value wouldn't change. They would own fewer shares, but each would be worth more.

Even more important was the education on insider trading. Most people don't realize that even seemingly innocuous acts like posting on social media or talking to family or friends about what's happening within the company can be illegal. The message we had to drill home was that the SEC looks into absolutely everyone who profits off stock trades in the run up to a major event at a company. A lot of people assume that the SEC won't make the connection between them and a friend or relative, but in reality, that's exactly what they're trained for. People had to understand, "If someone profits off what you know, the SEC will find out."

Of course, all of this entailed a massive increase in the workload for the finance team. Bob's team doubled in size just to handle everything, and the auditors started to visit every quarter. It's true that the IPO has changed many aspects of the company, financially speaking. However, the IPO—and all our financing—is a means to an end. The money ensures that we have the resources to get our treatments into the hands of doctors and patients. And if opportunities arise to create new treatments or drugs, we want to be ready to act.

That's why one of the best arguments for an IPO is that it makes it easier to raise money in the future. A company has to wait a year after holding the IPO, but at that point, with a minimal amount of documentation, it can sell more shares based on whatever the market price is that day. Morphic had around $300 million after the IPO, enough to help us push our first development candidates through the initial clinical phase while continuing to grow our development team and strengthen the partnerships with AbbVie and Janssen. Even with the additional employees and expenses related to development, our team has roughly a three-year cash runway.

On September 30th, 2019—almost three months to the day of the IPO—the entire Morphic team was invited to Times Square to celebrate as Praveen rang the bell to signal the close of the Nasdaq Stock Exchange. It was a momentous occasion. Tim stood beside the management team, and there were at least 50 members of the team on camera behind Praveen cheering him on. As Praveen pressed the button to close the exchange for the day, Morphic colored confetti drifted down from the rafters, photos flashed on the screens lining Times Square, and 4 networks carried the coverage live. To top it off, we spent the evening having dinner and plenty of wine in the city. For a few hours, amidst the clamor of NYC, we all relaxed.

Chapter 15

The Rocky, Regulated Road Ahead

As hiring accelerates and more people are brought on board at Morphic, attention at the senior level invariably is split. When we started the company, there were (at most) three projects to manage with a small, tightly-knit team. Now, there are more projects, more people, and less space available in our building. Those are all good problems to have, but they're certainly challenging.

There's no straightforward answer to meeting those challenges successfully. A young company simply doesn't have the established processes of a large, mature business. That's not necessarily a bad thing because it allows the team to figure out what works, rather than blindly following a set path. At Morphic, especially as we move toward clinical development, that lack of a roadmap can cause uncertainty amongst the team. We don't have a checklist to continuously refer to.

Much of the work ahead revolves around process, particularly when we reach development. FDA regulations aren't exactly open for interpretation, and everything we do has to be reportable. At this point, the focus will begin to shift toward executing specific processes and reducing the amount of variability in the work.

Setting the right priorities, putting the right people in charge of projects, delegation—it's all part of keeping a growing company on the road to success. Staying on that road is more difficult than many think. New, appealing paths begin to appear, just as the team begins to really hit its stride. The temptation, one that every biotech faces, is to begin focusing on targets in new areas. New opportunities often look better when they don't have a lot of data.

For instance, we've always had a focus and a deep understanding of one class of important biological receptors—the integrins. There are 24 integrins, as well as activators for each of them. In theory, that means we could have 48 different projects and combinations thereof. That's always been our competitive advantage. But there's also the temptation to take what we've learned from integrins and use it in other areas, pursue non-related targets. There is always a need to be doing something *different* or *better* to keep our competitive advantage, and it's a struggle to decide where to expand and when (if ever) to do it.

However, we're confident in our current course. By taking what we've learned from previous research, each new integrin project gets a little easier than the last. We're no longer in the stage where everything is fresh and exhilarating and all we can do is try to keep up. Now, we have money, we have projects, we have IP. Everything feels more stable.

But I have to say, even five years in, the company still excites me. So much has changed—and it's still constantly evolving. But that's the nature of success. As we approached the day we held our IPO, June 27, 2019, there had been a building sense of excitement. Management could say, "There's no guarantee this will happen," all they wanted, but the team knew it was more than likely to happen. Praveen was in New York the day we went public, watching the first trades, so we all waited until he came back to celebrate. The following day, celebration ensued. We got the team together, along with quite a few bottles of champagne, and Bruce started joyously speaking to the group about our accomplishments and how far we still had to go.

As he spoke, he kept popping champagne bottles—shooting corks in the general direction of the team. Corks were bouncing off the walls and ceiling, causing the team to duck and dodge with champagne in hand. You might say he was doing it to keep the team on their toes (luckily, no one was injured). After all, this was a momentous occasion, but not a final victory. The road ahead is still long, still arduous. Developing drugs can be exciting; it can be tedious. There will be more champagne corks, just as there will inevitably be setbacks and unexpected developments. To bring a biotech's dreams and drugs to fruition, everyone has to stay on their toes. Everyone has to stay sharp. And at Morphic, all of us will have to keep the two reasons why we built this company at the forefront of our minds—the patients and the science. Because in the end, the value of this company will not be measured by the price of its stock, but by the new treatments we will produce, and the patients in need who we will help.

Made in the USA
Middletown, DE
13 June 2020